HERITAGE DMUs

THE FINAL YEARS

ROGER MARKS

AMBERLEY

To Alex

First published 2015

Amberley Publishing
The Hill, Stroud
Gloucestershire, GL5 4EP

www.amberley-books.com

British Library Cataloguing in Publication Data.
A catalogue record for this book is available from the British Library.

ISBN 978 1 4456 4020 4 (print)
ISBN 978 1 4456 4029 7 (ebook)

Typesetting and Origination by Amberley Publishing.
Printed in Great Britain.

Introduction

Growing up with an interest in railways during the 1970s and 1980s, the Diesel Multiple Unit represented the plain vanilla side of the railway and I tended to shun them in favour of the richer pickings of locomotive-hauled trains. Familiarity bred contempt but the impending loss of something long taken for granted can sharpen one's appreciation, and so it was with the DMUs. With the 'Sprinter' revolution in full swing, mass withdrawal of first generation DMUs got underway; but British Rail also identified a need for a small fleet of certain DMU classes to be retained beyond their notional life span. These were somewhat cheekily given the official designation *heritage* units and while that probably didn't fool the travelling public, it perhaps went some way towards making me realise that I needed to take notice of these historic trains before they were gone for good. Little did I imagine at the time that not only would I devote much time to following and recording the final days of the heritage DMUs, but I would eventually end up driving the very last units in main-line service.

The mid-1990s brought the tornado of privatisation blowing through the state railway monolith and the heritage DMU fleet passed into the ownership of leasing company Angel Trains while the former Southern Region Diesel Electric Multiple Units became the property of Porterbrook Leasing. The passenger network was fragmented into twenty-six Train Operating Units, each of which was franchised to a commercial operator. Five of these new operators inherited leases on heritage DMUs and a sixth took on the operation of the equally antiquated DEMUs.

South Wales & West Railway (later rebranded as Wales & West) took over a handful of Class 101 and 117 units for branch-line services in Devon and Cornwall. Within a year of privatisation these had been displaced by more modern traction.

First Great Eastern took on a franchise that was facing a rolling stock crisis on the Marks Tey–Sudbury branch line. This was the only non-electrified part of the FGE network but no DMUs were available for lease. The BR-owned shadow franchise had subleased a Class 153 unit from Anglia Railways but the arrangement was terminated when Anglia could no longer spare any units. The solution came in the form of another sublease arrangement, this time with Silverlink Trains, who provided their three remaining Class 121 single-car units. The 121s worked on the branch for a little under two years until an Anglia unit was once again made available.

In addition to the above-mentioned 121s, Silverlink initially operated a fleet of Class 117 power twin sets on their non-electrified routes. These were replaced by ex-Central Trains Class 150 'Sprinter' units in 2000. The 121s lasted slightly longer but were also replaced by 150s the following year.

Scotland's railways were also left with a substantial fleet of heritage units upon privatisation. The new National Express-owned Scotrail franchise inherited eleven Class 101 power twin sets based at Corkerhill for use on Strathclyde PTE services, principally on the Glasgow Central–Whifflet and Glasgow Central–Paisley Canal services. Additionally, seven three-car Class 117 sets were based at Haymarket and used for Edinburgh–Markinch, Edinburgh–Cowdenbeath and Fife Circle services. The 101s were displaced by the summer of 2000 following a rolling stock cascade facilitated by delivery of the new Class 334 EMUs. Several sets were then transferred to Longsight TMD to bolster the remains of the First North Western fleet. The 117s were all withdrawn from service by the end of 1999, displaced by deliveries of new Class 170 units.

North Western Trains (rebranded First North Western after FirstGroup's buyout) inherited a substantial fleet of Met-Camm Class 101 units of various formations. They were used on a number of routes, especially around the Manchester area and on the North Wales Coast line and its branches. A steady withdrawal of the fleet coincided with the delivery of the new Alstom Class 175 'Coradia' units. The fifteen power-trailer sets were the first to go and by 2003 the original eleven power twin sets had been reduced to five, despite the arrival of additional ex-Scotrail units in 2000. The final workings took place in December 2003 and a farewell railtour was also run.

French company Connex won the right to operate the Network South Central franchise and thus took on the fleet of Class 205 and 207 DEMUs that were used on the non-electrified Ashford–Hastings and Oxted–Uckfield lines. Connex was later stripped of the franchise to be replaced by Govia, who operated under a temporary 'South Central' brand until sufficient upgrade work had taken place to relaunch the company as Southern. Part of this upgrade was the replacement of the DEMUs with Class 170 (later reclassified 171) Turbostars. The DEMUs bowed out in December 2004.

A small fleet of heritage DMUs also remained in service in various departmental functions, as test trains or rail head treatment trains. Their use declined throughout the first decade of the twenty-first century, and at the time of writing only two units remain in departmental service.

As 2002 drew to an end, many assumed that the end was nigh for heritage DMUs in main-line passenger service. The writing was already on the wall for First North Western's Class 101 fleet, which was now only being used to cover three daily diagrams. South Central was still soldiering on with its fleet of DEMUs but the replacement fleet of Adtranz Turbostars was in the pipeline and would start to appear within a year. Then came an unexpected announcement from Chiltern Railways; a Class 121 'bubble car' was to be refurbished for passenger service to operate peak-time services on the Aylesbury–Princes Risborough branch. This was a bold plan to release a Class 165 unit to strengthen peak-time commuter services on the main line. Fully upgraded for twenty-first century use, 121020 entered service in May 2003 and proved to be a success. In 2011 it was joined by a second unit.

Arriva Trains Wales followed Chiltern's lead in 2006 when it purchased Class 121 single-car unit 55032 out of preservation and put it into service on the short Cardiff Bay branch. Despite the less than taxing workload, reliability proved to be a major problem and the unit was withdrawn from service in 2013, leaving Chiltern Railways as the sole passenger heritage DMU operator.

The obvious question to ask at this point is: how much longer will the Chiltern 121s continue to run? They are, after all, both well on their way to their seventieth birthdays. The absolute deadline for withdrawal is 1 January 2020, when the Rail Vehicle Accessibility Regulations stipulate that all trains must be fully accessible. The 121s come nowhere near to meeting the specification and modifications are not viable. Whether or not they continue running up until the end of the 2019 summer timetable will depend on a number of factors, including their general reliability and the availability of spares. Whenever the curtain does finally fall on the Chiltern units, it is worth remembering that heritage DMU operation was supposed to have ended over a decade ago and every year since has been a bonus. Fortunately, heritage DMUs have proved to be a valuable asset to the preservation movement so it will be possible to enjoy DMU traction on Britain's heritage railways for many years to come.

Roger Marks
October 2014

First North Western

At the beginning of the privatisation era, North Western Trains inherited an all Met-Camm fleet of power-twin and power-trailer Class 101 sets, as well as a pair of three-car sets. Power-trailer set 101658 (51175/54091) is seen here at Rose Hill with a service for Manchester Piccadilly on 6 March 2000, by which time the company had been rebranded as First North Western. The unit nevertheless still carries full BR Regional Railways livery.

The First North Western 101s were used extensively on the North Wales Coast line and its branches. Having arrived with a shuttle service from Llandudno, power-twin set 101680 (53204/53163) stands in the bay platform at Llandudno Junction on 7 March 2000.

On Sunday 5 March 2000, one of Longsight's power-trailer sets, 101662 (53228/54055), is stabled for the weekend at Chester in the company of younger brethren in the form of 153332 and 158754.

Power-twin set 101681 (51228/51506) passes the magnificent backdrop of Conwy Castle, working the 2D66 Llandudno–Holyhead service on 8 May 2000. (Chris Booth)

A conference of different-liveried 101s at Manchester Piccadilly on 8 November 2002. Strathclyde orange 101693 is on the left, with Caledonian blue 101692 and British Railways green 101685 on the right.

Green-liveried 101685 entered the post-BR era as a three-car set but soon lost its centre trailer. Reduced to a two-car power-twin set (53164/53160), it was photographed here at Sheffield with a Hope Valley line service for Manchester Piccadilly on 19 April 2001. No. 101685 was soon nicknamed 'Daisy' owing to its livery and resulting similarity with the *Thomas the Tank Engine* character of that name.

In addition to the core fleet of 101s, Longsight depot also operated a pair of Class 101 power-twin sets that had been transferred from the former Network South East sector. They differed from the rest of the FNW fleet in that they retained their former first-class seating (albeit declassified) and some standard seating had been removed and replaced by luggage racks when they were in use on the Reading–Gatwick service. No. 101835 (51498/51432) was repainted into Regional Railways livery but retained its Reading depot set number, L835, in lieu of the full TOPS number. The unit is seen here at Sheffield forming a service for New Mills on 16 June 1999. (Chris Booth)

The other ex-Network South East set, 101840 (53311/53322), even retained its NSE livery and branding in addition to the Reading depot set number, L840. It is pictured here at Bidston, Merseyside, on 8 March 2000, waiting to form a service to Wrexham.

After the demise of Class 101s on Scotrail, some units were reallocated to Longsight for use by First North Western. Still carrying the unique blue Strathclyde livery it was given in Scotland, set 101692 (53253/53170) stands under the overall roof at Manchester Piccadilly alongside 156459 on 8 November 2002.

No. 101678 (51210/53746) calls at Romiley with a Manchester Piccadilly–Rose Hill service on 3 July 2002. FirstGroup tend to apply their logo to everything that moves with an almost evangelical zeal, but the 101s all remained in the liveries they carried when they were inherited, logos and all. Clearly First were not keen to be too closely associated with these ageing units.

No. 101676 (51205/51803) stands at Manchester Piccadilly on 30 May 2002. DMSL 51803 shows a livery variation that was unique to this vehicle; the light blue stripe was carried right to the cab front instead of changing to three dark blue lines at the start of the cab door.

On a dull winter's afternoon on 16 December 2003, just days before withdrawal, 101676 rolls into Romiley, Greater Manchester, with a service from Manchester Piccadilly to Marple via Bredbury. Romiley was a good location to see 101s in their last days as it was served by trains on both the Manchester–Guide Bridge–Rose Hill and the Manchester–Bredbury–Marple routes.

With the guard keeping a watchful eye from the window, 101676 accelerates away from Hyde North with a service for Manchester Piccadilly from Rose Hill. DMBS 51205 is recognisable by its offset headlight in this photograph taken on 28 November 2003.

No. 101678 crosses Hyde Junction and rounds the curve into Hyde North station with a service for Rose Hill on 28 November 2003. Note the evidence of patch painting across the top of the unit's front end.

Having shunted across to the Down platform via the crossover in the background, 101678 is ready to form a service for Manchester Piccadilly at Marple on 28 November 2003. Prominent on the roof is the antenna for the NRN radio system.

Left: Carrying full Strathclyde Passenger Transport orange livery, 101693 (51192/53266) stands at Manchester Piccadilly on 8 November 2002, while a fitter attends to a faulty headlight. Unusually by this time, the leading vehicle retains a windscreen wiper on the secondman's side.

Below: An extremely battered and rusty 101680 stands on the stop blocks at Manchester Piccadilly as passengers trudge down the platform from a Class 323-formed service that has arrived on the same platform on 16 December 2003.

No. 101685 stands at Manchester Piccadilly with a service for Rose Hill on 16 December 2003. The passengers are loaded, the doors are closed; all that is needed is a driver. By this time a mere five units remained in service with FNW, of which three were needed per day to operate the Class 101 diagrams.

Having just departed from Marple, 101676 heads smokily towards Marple North Tunnel beneath the Peak Forest Canal on one of the very last Class 101 workings, the 2H69 12.50 Marple–Manchester Piccadilly service on 22 December 2003. (Chris Booth)

On 21 December 2003 First North Western ran the 'Final Farewell' tour to mark the withdrawal of the Class 101 fleet from passenger service. The formation for the railtour (headcode 1T01) consisted of three two-car units: Regional Railways-liveried 101676, Strathclyde orange-liveried 101693 and BR green-liveried 101685. Note that the latter has been adorned with *Daisy* nameplates. The tour visited various parts of the FNW network and is seen here at Preston. (Charlie Hulme)

The 'Final Farewell' tour in the midday winter sunshine at Buxton on 21 December 2003. Note the tour headboard adorning the careworn front end of 101676. (Charlie Hulme)
Inset: A ticket from the 'Final Farewell' tour. (Charlie Hulme)

No. 101685 *Daisy* at the lightly used Lancashire terminus of Heysham Port on 21 December 2003 on one of the more unusual legs of the tour. (Charlie Hulme)

Towards the end of the tour, 101676 is photographed at Barrow-in-Furness on 21 December 2003. The sun has symbolically set on the 101s and the final couple of hours of the journey on 21 December 2003 are made in darkness. (Charlie Hulme)

One of the driver's desks on FNW unit 101678. The vast majority of heritage DMUs share the same basic controls, but the arrangement of those controls differs from class to class. A distinctive feature of the Class 101 cab is the splitting of the gauges into two separate clusters, with the speedometer and tachometer directly in front of the driver and the duplex (vacuum) and control air gauges offset to the right by the AWS indicator.

The desk on the secondman's side of 101678. Prominent are the handbrake wheel and the emergency brake valve. Also present are the heating and lighting controls, and a windscreen wiper switch, redundant by this time on the FNW 101s. At the time of this photograph the TPWS equipment had not yet been fitted; it would be attached to the side wall of the cab (far right).

The guard's compartment of DMBS vehicle 51210 of unit 101678, with DMSL vehicle 53746 visible through the corridor connection. The objects attached to the end wall on the right are track circuit operating clips.

The saloon interior of DMSL vehicle 53746 of unit 101678, looking towards the cab. The seats are covered with standard Regional Railways moquette.

The forward part of the passenger saloon in DMBS 51205 of unit 101676. This particular vehicle was finished internally with green Formica in the saloons. On other vehicles the Formica was cream or cream/dark blue.

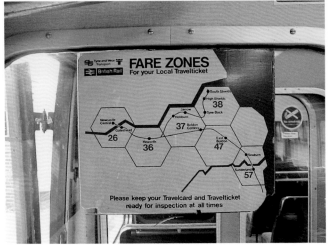

A sticker on one of the vestibule partitions in DMBS 51205 of unit 101676 tells of a previous life in Tyne & Wear.

Scotrail

No. 101687 (51247/51512) at Glasgow Central with EMU 303003 on 15 May 1997. Both units carry the orange livery of Strathclyde Passenger Transport, with a small logo on the cab front signifying that they are operated by the former National Express-owned Scotrail franchise.

An unidentified Class 101 set departs from Glasgow Central on 13 May 1997, past an enthusiast who seems more interested in the photographer than the train.

In 1996 Scotrail/Strathclyde Passenger Transport outshopped 101962 (53170/53253) in a one-off livery based on Caledonian Railway blue to mark the start of a new service between Motherwell and Cumbernauld. The unit is seen here departing from Glasgow Central with a service for Whifflet on 14 March 2000. (Chris Booth)

Seven ex-Tyseley three-car Class 117 sets were transferred to Haymarket in 1993/94, where they were primarily used to work the Fife Circle services. Viewed from the top of the Wallace Monument, an unidentified set departs from Edinburgh Waverley in June 1999. At the end of the summer the remaining sets were reduced to two-car formation and spent their last couple of months running as four-car (2+2) formations. By November 1999 sufficient Class 170 units had been delivered to allow withdrawal of the 117s. (Lewis Bevan)

Following its withrawal from Scotrail in 1999, 117306 (51369/51411) was purchased by Chiltern Railways as a source of spares for the fleet of departmental Sandite units that they ran on behalf of Railtrack. DMBS 51369 was stripped for spares and scrapped within eighteen months of arriving at Aylesbury but DMS 51411 was retained and used as a storage facility for DMU spares until the summer of 2009. It is seen here awaiting asbestos stripping on 11 September 2009.

An interior view of DMS 51376 in 2003, when it was running on the West Somerset Railway. It is substantially in as-withdrawn condition with the exception of the primrose yellow paint on the window frames and doors, which were previously painted tan and off-white respectively.

The derelict and partially stripped interior of DMS 51411 at Aylesbury in September 2009. Having spent the previous eight years as a parts store, it was awaiting scrapping when this picture was taken.

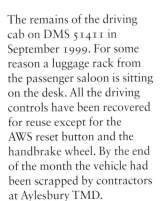

The remains of the driving cab on DMS 51411 in September 1999. For some reason a luggage rack from the passenger saloon is sitting on the desk. All the driving controls have been recovered for reuse except for the AWS reset button and the handbrake wheel. By the end of the month the vehicle had been scrapped by contractors at Aylesbury TMD.

DMBS 51369 from 117306
in the process of being cut up
by contractors at Aylesbury
TMD on 29 May 2002.

After withdrawal by Scotrail, 117311 (51352/51376), along with centre TSL 59505 from 117301, was purchased by 275 Railway Squadron, Territorial Army. The unit was initially based on the West Somerset Railway, where it was used on scheduled passenger services. It is seen here at Williton on 8 April 2003.

In 2007 the Army's Class 117 set was moved to the former MOD base at Long Marston, where it continues to be used for training exercises. With DMBS 51352 leading, the unit is seen at the Warwickshire base on 17 September 2011.

First Great Eastern

As mentioned in the introduction, First Great Eastern subleased three Class 121 single-car units from Silverlink Trains in 1997 to operate the Marks Tey–Sudbury branch line. Nos. 121027, 121029 and 121031, still carrying obsolete Network South East livery, were duly reallocated to Ilford depot. They later received First Great Eastern vinyls and 'Sudbury Line' route branding. No. 121027 is seen here at the Sudbury terminus with a service for Marks Tey. (Martin Hawkes)

No. 121029 at Marks Tey with an FGE service for Sudbury on 31 December 1997. It still carries its NSE route branding for the Bedford–Bletchley line and FGE vinyls have yet to be applied. The 'bubble cars' did not stay in Essex for long and all three had returned to Bletchley by 1999. (Ian Buck)

Silverlink Trains

Silverlink Trains inherited a fair-sized fleet of Pressed Steel Class 117 and 121 DMUs from British Rail. The DMUs were used to operate Silverlink's three non-electrified routes: Bletchley–Bedford, Gospel Oak–Barking and Clapham Junction–Willesden Junction. No. 117706 (51366/51408) is seen departing from the Marston Vale bay platform at Bedford with a Silverlink service for Bletchley on 27 April 1999.

Class 117 two-car set L720 (51354/51396) is seen here arriving at Clapham Junction with a service from Willesden Junction. The Class 117 sets all remained in Network South East livery until withdrawal.

Class 117 power twin-set L702 (51356/51398) stands at Bedford with a Marston Vale line service for Bletchley. Behind the unit is Sandite car 930079 (ADB 977579), a former EMU vehicle.

Nos 121027 and 121029 depart from Millbrook with the 15.30 Bletchley–Bedford service on 14 October 1998. In order to aid revenue protection, off-peak services generally ran with only one unit available for passenger use. However, the presence of the second unit provided much needed insurance against train failure. (Lewis Bevan)

Nos 121027 *Bletchley TMD* and 121031 *Leslie Crabbe* pass Forders Siding en route for Bedford on 6 July 1999. No. 121027 had recently been repainted into a version of Silverlink's livery. (Chris Booth)

Above: Detail of 121029 (55029) *Marston Vale* and 121031 (55031) *Leslie Crabbe*, coupled together by their no. 2 ends at Bletchley on 20 August 1999. Both units carry the abstract logo of Silverlink Trains and 121029 also retains its NSE route branding crest. (Lewis Bevan)

Left: Front end detail of 'bubble car' 121027 (55027) in the slightly garish Silverlink livery at Bletchley on 27 March 1999. No. 121029 later received this livery also, but only briefly before withdrawal. The nickname 'bubble car' was coined in the 1960s, allegedly because single-car units had windows on all sides and, in railway terms, they were very small trains.

The fourth Class 121 unit in the Silverlink stable was green-liveried L123 (55023), which was withdrawn in 1997. This is the interior of the unit, showing the partition that once separated the smoking and non-smoking saloons. By the time this photograph was taken in 2005, the unit had spent several years in preservation at the Chinnor & Princes Railway but the interior still remained in much the same condition it had been in when last in service, with NSE moquette on the seats.

The nameplate of of 121031, cast in the house style of the former North London Railways TOU. The plates were attached to 121031 in 1999, and had previously been carried by Class 117 DMBSO 51358 from unit 117705. The *Marston Vale* nameplates on 121029 were cast in the same style and had previously been carried by Class 117 DMBSO 51332 from unit 117700. No. 121027 received brand-new Silverlink-style *Bletchley TMD* nameplates in 1999. After the 121s were withdrawn, all three names were transferred to Class 150 units.

Chiltern Railways

In a surprise move, Chiltern Railways announced in late 2002 that they were intending to refurbish a Class 121 single-car unit for passenger service. The intention was to use it on the Aylesbury–Princes Risborough branch line during peak times, thus releasing a Class 165 unit for strengthening a main-line service. Former Sandite and route learning unit 960002 (55020) was transported to LH Group Services, Burton-on-Trent, where it was overhauled and restored to passenger configuration as 121020. It is seen here at Princes Risborough on its delivery run, partnered with 960010 for added insurance, on 2 April 2003.

Right: As they already operated railhead treatment trains for Railtrack, Chiltern had a ready-trained pool of heritage DMU drivers although a handful of guards required training. More in need of training were the passengers, some of whom had never even seen a slam door train, the species having been extinct on the Chiltern Line for over a decade. Now fully settled into service, 121020 passes Askett, midway between Monks Risborough and Little Kimble, with an Aylesbury-bound service on 11 May 2006.

Below: In an attempt to make the unit look slightly less old fashioned, the decision was taken to reroute the exhaust pipes internally when it was refurbished. One of the holes through which the exhaust pipes were originally routed can be seen in the buffer beam of 121020 as it leaves Aylesbury station, bound for the Traction Maintenance Depot. The overall Chiltern blue livery with silver lining was chosen because it suited the lines of the heritage unit better than a straightforward copy of the livery used on Chiltern's Class 165 and 168 units.

No. 121020 awaits its next afternoon peak working to Princes Risborough at Aylesbury's platform 1. The regulations that were drawn up to allow a non-compliant heritage unit to re-enter passenger service are tight and they specify that only this platform at Aylesbury and the bay platform at Princes Risborough are permitted to be used in normal passenger service. Route learning unit 960014 is standing in the adjacent branch siding in this photograph from 9 June 2006.

Inside Aylesbury TMD, 121020 awaits an A Exam on the south end of 'C' road on 11 February 2014. The Chiltern blue livery on this unit was over a decade old when this picture was taken and close examination will reveal several places where it has been touched up by patch painting.

For many years, special Quaintonian charter trains have operated on selected bank holidays between Aylesbury and the Buckinghamshire Railway Centre at Quainton Road. The requested motive power switched from Class 165 to Class 121 as soon as 121020 became available. Here, 121020 passes the former Waddesdon station (closed 1936) with an Aylesbury-bound Quaintonian service on 7 May 2007.

On Saturday 4 September 2010, Chiltern Railways celebrated the fiftieth birthday of 121020. The branch service to Princes Risborough, normally booked for a 165 on a Saturday, was covered by 121020, and various forms of entertainment were laid on at Aylesbury. On static display was Wrexham & Shropshire Class 67 locomotive 67015 *David J. Lloyd*. *Inset*: One of the special One Day Bubblecard tickets that were produced for 121020's fiftieth birthday.

During 2010, Chiltern Railways identified the need for a second Class 121 unit in order to bolster availability. Preserved unit 55034 was purchased from Tyseley Locomotive Works, who were also given the contract to refurbish it for main-line passenger use. It is seen here on its delivery run passing Rushbeds Wood, with Brill Tunnel visible in the background, on 4 May 2011.

No. 121034 approaches Marsh Crossing with the 2A57 18.00 Princes Risborough–Aylesbury service on 30 April 2013. The livery chosen for this second unit was BR green with small yellow panels, an option that had been considered but rejected for 121020 in 2002.

Right: A close-up of the no. 2 end of 121034 on 11 July 2011 in the tamper siding at Aylesbury, where the 'bubble cars' are normally stabled. Unlike 121020, the exhaust pipes on this unit were not rerouted internally, so it retains its distinctive 'antlers'. It also carries its full unit number on the no. 2 end, but only the last three digits on the no. 1 end.

Below: Seen from the high vantage point of the Bourg Walk footbridge, 121034 stands at Aylesbury's platform 1 with a service for Princes Risborough. Chiltern-owned departmental units 960014 and 960301 are stabled in the branch siding alongside in this photograph from 3 May 2013.

No. 121034 stands in the bay platform at Princes Risborough with the 2A18 10.00 service for Aylesbury on 14 April 2014. Note the unlikely destination displayed on the roller blind. While 121020 was fitted with an electronic destination display when it was refurbished, 121034 retains its original roller blind, complete with a set of unhelpful Tyseley-based destinations.

No. 121034 undergoing routine overnight maintenance in 'C' road at Aylesbury TMD on 22 October 2013. The extensions to the bottoms of the passenger and guard's doors, which overlap the footstep below, are visible here. This is part of the electromagnetic secondary door locking system, which will be covered in more detail later.

The quintessential country railway: 121034 calls at Little Kimble with the 2P75 17.30 Aylesbury–Princes Risborough service on 30 April 2013. The former station building now serves as a private dwelling. Above the station is Ellesborough church, and perched atop the Chiltern escarpment is the Coombe Hill Boer War monument. Little Kimble is the nearest station to Chequers and in 1998 it was visited by the royal train, dropping off the wives of the heads of state for a reception at the Prime Minister's country residence.

No. 121034 made its debut on the Quaintonian service on 29 August 2011. With Quainton Road situated on the horizon, the green 'bubble car' is passing the former Waddesdon station, bound for Aylesbury, despite what the destination blind would have you believe.

The Quaintonian was in the hands of 121034 again on 6 May 2013. It is seen here in the picturesque setting of Quainton Road. Only the former Up line remains through the station and the Network Rail metals have no physical connection with the Centre's Up and Down yards on either side.

Having finished its Quaintonian duties for the day, 121034 heads back to Aylesbury TMD on 6 May 2013 to await the next morning's commuter workings. Note the marker/tail lamps, which have been replaced by modern LED fittings.

Detail of the interior of the guard's compartment on 121020, showing one of the rerouted exhaust pipes. In winter they provide the guard with some welcome additional heating. In summer their presence is less welcome. The right-hand window is not original and was added when the unit was converted for route learning use.

The guard's area in 121020. On the left is the emergency brake valve and gauge, while the bulkhead behind the seat houses the train's heating and lighting controls and the PA system.

The guard's compartment of 121034 is arranged differently to that of 121020 and an enclosed cupboard is provided for equipment storage. The emergency brake valve and gauge can just be seen on the extreme left. Other differences compared with 121020 are the fluorescent lighting and the fact that the electromagnetic secondary door locking system is fitted to the guard's doors (the white protrusion at the top of each door).

The interior of 121020. The 2003 refurbishment included the replacement of the floor and all internal panelling. Even the wooden window frames had to be replaced in metal to comply with modern fire resistance standards. The double windows adjacent to the second row of seats from the camera position identify the location of the original partition that divided the saloon into smoking and non-smoking sections. An ex-Class 165 Passenger Information System has been added, visible above the internal cab door. Over half the original luggage racks have been omitted; the remainder, which have been powder coated, are more than sufficient for the unit's current workings.

By way of contrast, this is the interior of 121034. On both 121s the seats have been upholstered in the same moquette as the Chiltern 165s. However, 121034 also sports Class 165 style grab handles. The luggage rack capacity has been reduced in the same way as it has on 121020, but in this case the remaining racks retain their alloy finish. Unlike its sister, this unit has not been fitted with PIS.

Comparative cab views of the two Chiltern Railways passenger Class 121 units. No. 121034 is on the left and 121020 on the right. These photographs were taken before the units were fitted with GSM-R radio equipment.

Both Chiltern passenger 121s were fitted with forward-facing CCTV equipment in the early months of 2014. The cameras have been mounted on the driver's desks, in front of the centre windscreen. This is one of the camera housings on 121020.

Detail of the pass-comm apparatus (communication cord in old money) in the no. 1 cab of 121020. The chain that disappears into the bulkhead (top right) runs the length of the saloon and can be accessed by passengers at strategic points along its length. When a passenger pulls the cord, it moves the lever at the top and operates a valve inside the red painted casting, allowing in air through the holes in the side. This destroys the vacuum in the brake pipe and applies the brake. On the left is a microswitch, a modern addition, fitted only to the three 121s that were upgraded for twenty-first-century service. Its function is to release the electromagnetic secondary door locking system when the pass-comm is pulled.

One of the guard's control panels on 121034. It consists of a light switch, driver/guard communication buzzer and the locking button for the electromagnetic secondary door locking (the unlocking is performed by the driver).

Detail of the electromagnetic secondary door locking mechanism on one of 121020's saloon doors. An extension on the bottom of the door holds a metal block that fits against an electromagnet in a recess below the doorstep. When the electromagnet is energised, it holds the door closed.

Detail of one of the British Railways crests on 121034. However dusty the unit gets, the crests are usually kept polished up.

An unofficial bodyside embellishment on 121020 demonstrates that not everyone at Aylesbury holds the bubble cars in high regard.

Arriva Trains Wales

Arriva Trains Wales purchased Class 121 'bubble car' 121032 (55032) in 2006 for use on the Cardiff Bay shuttle service. No. 55032 had been a Midland Region unit and finished its days as a Sandite/route learning vehicle. The conversion process for this role was not anything like as extensive as that applied to sister units on the Western Region and the unit remained substantially in as-built condition until it was purchased by Pete Waterman for preservation in the early 1990s. It is seen here on display at the Crewe Basford Hall open day on 28 August 1995.

Upon purchase by ATW, 55032 was overhauled at LNWR Crewe, refurbished internally and fitted with electromagnetic secondary door locking. It entered service on the short Cardiff Bay shuttle service in August 2006. Towards the end of its days in Cardiff, 121032 is seen crossing the bridge over the walkway between Bute Street and Lloyd George Avenue in Cardiff with 2B62, the 12.48 Cardiff Queen Street–Cardiff Bay service on Saturday 2 February 2013. (Dave Evans)

Having suffered from generally poor availability in Cardiff, 121032 was withdrawn from service in March 2013. It was subsequently purchased by Chiltern Railways and after a period of storage at Tyseley it was moved to Aylesbury in the early part of the following year, where it is seen here in the branch siding on 13 April 2014. At the time of writing (October 2014), Chiltern Railways had not announced their plans for this unit, but it may be converted into a route learning vehicle to replace 960014 (55022).

Above: The saloon interior of 121032. The seats carry the standard Arriva Trains Wales moquette and the doors, window frames and end partitions are finished in a pale blue. Like the Chiltern units, the partition that separated the smoking and non-smoking has been removed. CCTV has also been fitted. Only four sets of passenger doors on each side are fitted with secondary locking, the remainder being sealed up.

Right: The no. 2 end cab of 121032. The cabs were not refurbished to anything like the same standard as the Chiltern 121s. No door controls are fitted as this is the guard's responsibility on ATW. The only significant additions since the unit's BR days are the TPWS panel, GSM-R equipment and a pair of brake cylinder Low Topside warning lights, situated between the driver's brake valve and the AWS indicator.

South Central

The Network South Central franchise was let to Connex in 1996 and the new private operator inherited a fleet of Class 205 and 207 DEMUs to work their remaining two unelectrified lines: Hastings–Ashford and Oxted–Uckfield. On the latter route, the first of the class, 205001 (the former 1101), calls at Edenbridge Town with an Up service on 10 July 2003, by which time Connex had lost the franchise to Govia. This unit survived right to the end of Class 205 operation in December 2004 and subsequently went for preservation at the East Kent Railway.

By the turn of the twenty-first century, Selhurst depot retained a mixed bag of Class 205 and 207 units, drawn from various batches and subclasses. No. 205025 is a unit from the third batch of Class 205 units and was constructed in 1959. It is seen here at Rye with an Ashford–Hastings 'Marshlink' service on 5 September 2002, just over a year after Govia had taken over the franchise.

A final passenger boards 205012 at Cowden before it departs with an Uckfield–Oxted service on 23 January 2003. This unit was scrapped in 2005, one of the few of the twenty-first-century DEMU survivors that did not find a place in preservation. The Southern Region DEMUs were nicknamed 'Thumpers' by enthusiasts owing to their distinctive engine sound.

No. 205032 departs from East Croydon in the evening sunlight on 24 June 2002, bound for Selhurst depot. It had previously formed an Uckfield–East Croydon service. The smaller route indicator panel identifies this as one of the final batch of the class, delivered in 1962 for services in Berkshire. After withdrawal in 2004, this unit moved to the Dartmoor Railway.

Having alighted from an East Grinstead-bound EMU on 24 January 2004, passengers wait at Oxted's bay platform to board 205033, which will shortly depart for Uckfield. No. 205033 survived right to the end of Class 205 operation in December 2004 and then went on to a new life in preservation on the Lavender Line.

Following withdrawal, TSO 60677 of 205032 is seen in heavily vandalised condition while in storage at Meldon Quarry in Devon on 29 July 2007.

No. 207017 starts away from Appledore with a 'Marshlink' service bound for Hastings on 18 February 2004. The unit retains its de-branded Connex livery, a situation that would apply to the whole Class 205/207 fleet until their withdrawal. At the time of writing, this unit is preserved at the Spa Valley Railway.

No. 207203 is seen here at Hastings, forming a service for Ashford on 5 September 2002. Three Class 207/2 sets were refurbished internally in 1991 and fitted with gangway connections. Each unit later had an ex-Class 411 EMU TSOL inserted as a centre car. As can be seen here, the TSOL is visibly wider than the two Class 207 vehicles that enclose it.

No. 207202 *Brighton Royal Pavilion* powers away from Cowden with a service for Uckfield on 28 January 2004. On the right, under a light dusting of snow, is the trackbed of the former Up line and the disused Up platform.

No. 207202 in the bay platform at Oxted on 28 January 2004. Note the original class designation, 3D, stencilled onto the unit front. No. 207202 was withdrawn from service just over six months later but found a home in preservation on the East Lancashire Railway.

One of the attractive *Brighton Royal Pavilion* nameplates that were applied to 207202 in 2005. No. 207201 was named *Ashford Fayre* at the same time. The cast nameplates were carried on the DMBSO vehicles.

With a Class 423 EMU standing in the Down/bidirectional platform, 207203 waits to perform a shunt over to the bay platform, where it will form the next service for Uckfield. This unit, seen on 24 January 2004 minus its ex-EMU centre car, was later preserved on the Swindon & Cricklade Railway.

To cover for South Central unit shortages, a Class 201 unit was hired from time to time from Hastings Diesels Ltd. DMBSO 60000 *Hastings* stands at Ashford International on 18 February 2004, part of a three-car power-trailer-power formation that will shortly work back to Hastings.

DMBSO 60016 *Mountfield* leads the same three-car formation into Appledore with a Hastings–Ashford service on 18 February 2004. This DMBSO carried a full yellow end in contrast to the small yellow panel applied to 60000.

Hastings Diesels Ltd still operate a main-line-certified Class 201 unit, the formation of which can be varied to suit demand. In addition to South Central, the unit has been hired to various TOCs over the years, including Anglia Railways, Valley Lines and Wales & West. However its main bread and butter is railtour work. DMBSO 60116 *Mountfield* is seen on one such duty, at the head of a six-car formation approaching Quainton Road on 14 May 2011.

Above: An interior view on board 205032, upholstered in NSE moquette and similar in layout to many contemporary classes of Southern Region EMU. The lack of gangway connections or CCTV rendered these units susceptible to interior vandalism, although this particular compartment seems to have escaped unscathed.

Left: A toilet compartment in unit 205033. Yellow Formica and yellow ceramic toiletware recall the early 1960s origin of the unit; the only concession to modernity is the paper towel dispenser.

No. 205205 was extensively rebuilt in 1979–80 with a completely revised interior and various technical modifications. Gone is any trace of varnished wood. The seat back luggage racks have been replaced by wall-mounted ones; fluorescent lighting has been fitted; and gangway connections have been installed. However, funding was not available to roll out the modifications on the remainder of the fleet, so 205205 became the sole member of the 205/1 subclass.

The interior of one of the standard-class saloons on 207017. The only apparent modification from original specification is the NSE moquette on the seats, although for some reason the back panel of the bench seat at the end retains its original BR pattern fabric.

Inside the TSOL vehicle, 70549, in unit 207202. This is not the vehicle's original interior, as it was stripped and refitted at Swindon Works in the 1980s, when it formed part of a Class 411 unit.

The interior of DTSO 60904 of unit 207202. NSE blue moquette gives the interior a relatively modern look, although the hair grease stains are somewhat off-putting. Fluorescent lighting has been fitted in place of the original tungsten bulbs.

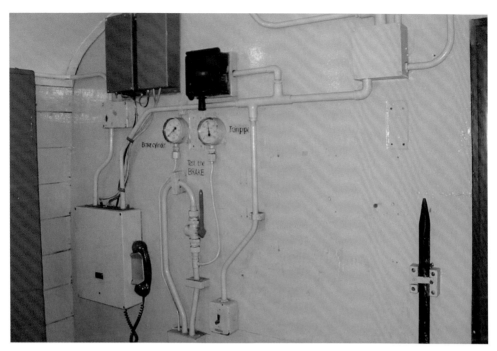

A corner of the brake van compartment in DMBSO 60127, part of unit 207203, showing the guard's brake gauges.

The nicely restored interior of Class 201 DMBSO 60000 *Hastings*. Tungsten lighting, varnished British Empire woods and seats deep enough to half-disappear within – a delight!

The cab interior on DMBSO vehicle 60154 of unit 205001. Prominent in the centre is the pedestal for the Cab Secure Radio system.

A close-up of the driver's desk on DTSO 60916 of unit 207017.

Departmental Units

Laboratory 19 Iris 2 was formed of two former Class 101 vehicles, DMBS 53222 (977693) and DMCL 53338 (977694), converted for departmental use in 1991. It was principally used for radio survey work, although towards the end of its working life it was also equipped with video survey equipment. The unit is seen here on display at the Old Oak Common open weekend on 6 August 2000, at which time it carried the red and grey livery of Serco Railtest. It was later repainted in Railtrack lime green and blue and for its last few years prior to withdrawal in 2007 it carried Network Rail yellow. It has since been preserved and at the time of writing resides at the Barry Tourist Railway.

Former Class 100 DTCL 56106 was withdrawn from passenger service in 1983 and entered departmental use as part of the Crewe works test train, where it was numbered ADB977191. Having been stored out of use for many years in Basford Hall yard, it survived into the post-BR era and is seen here at Adtranz Crewe in an advanced state of decay on 21 May 2000, shortly before it was scrapped.

No. 79900 was one of two similar single-car Derby Lightweight units that entered service in 1956. Having spent most of its life working between Banbury, Verney Junction and Bletchley, it was withdrawn from passenger service in 1967 and transferred to departmental stock as *Test Car Iris*. Numbered 975010, its main task was to undertake radio survey work. For the last few years of its working life the unit lost its research department colours in favour of a repaint into original green, in which state it is seen here at London Waterloo on 1 July 1998. Around two years later it was withdrawn from service and sold for preservation.

Verney Junction, 1965? Well, it could almost be! In fact this is Swanwick Junction on the Midland Railway, Butterley, and 79900 has been beautifully restored back to original passenger configuration in this photograph from 15 July 2006.

Following withdrawal from Silverlink service, 121029 (55029) was converted into a video survey and track inspection unit for Eurailscout GB, numbered 977968. It is seen here at Crewe on 30 March 2004. Most of the bodyside doors have been panelled over, giving the unit a very different appearance. It later worked as a route learning unit for Carillion Rail before finding a home in preservation at the Rushden, Higham & Wellingborough Railway. (Rowan Cranshaw)

Network South East converted seven of their Class 121 units into route learning and Sandite units during the early 1990s. Initially based at Reading TMD, several were later moved to Aylesbury, where 960002 (977722/55020) is seen here on 29 July 2000, stored out of use. This unit was later chosen for reconversion to passenger stock, becoming Chiltern Railways' 121020.

Another of the NSE route learning and Sandite conversions was 960012 (977873/55022). By the time it was photographed here at Aylesbury on 29 July 2000 it was being operated by Chiltern Railways but it was later sold to South West Trains.

South West Trains used 960012 as a route learning unit. It received a version of the SWT livery and was given the name *John Cameron*. In 2009 it was sold to the Swanage Railway for preservation, where it was overhauled and returned to original BR green livery. It is seen here passing Holes Bay with the 5Z60 Bournemouth Traincare Depot–Swanage ECS working, the unit's delivery run to the Swanage Railway on 21 March 2009. (Antony Henley)

Class 122 unit 55019 entered departmental service at Bletchley as a route learning and Sandite unit under the auspices of the former Network South East sector, becoming the only Class 122 DMBS to carry that sector's livery. It was numbered ADB975042 in the departmental series. By the end of the twentieth century it had been allocated to Aylesbury and carried the grey, white and tan livery of Railtrack plc. It is seen here at Aylesbury TMD on 20 September 2000.

No. 55019 was given a coat of Network Rail yellow paint during 2004, the specification for which included red painted buffers. It is pictured here in the branch siding at Aylesbury on 5 June 2005. By this time it had been given the departmental unit number 960015.

Having spent several years in storage, a faded and graffiti-covered 960015 stands out of use in the branch siding at Aylesbury on 20 June 2011. By this time Chiltern Railways had taken ownership of the unit and in 2012 they sold it to the Llanelli & Mynydd Mawr Railway for the nominal sum of £1.

A view of the no. 2 cab of 960015. This unit, unusually, has its Cab Secure Radio console attached to the driver's desk. Most other units had it fixed to the bulkhead next to the driver's seat. Another interesting feature of this vehicle is that it is equipped with the obsolete GWR/Western Region ATC system.

The no. 2 end of 960015 seen from the route learning saloon. The partition behind the driver's seat differs from those found in the departmental Class 121 units. Maroon unit 960010 can be seen through the windscreens.

No. 55019 had not been put through the British Rail DMU refurbishment programme of the 1970s and so as 960015 it retained its distinctive tungsten light fittings, along with the requisite splattering of Sandite paste.

Two of the Sandite hoppers inside 960015 (55019). Sandite is used as a generic term for various types of traction-improving pastes containing sand and ferrous particles. It is applied to the railheads to aid adhesion and track circuit activation during the leaf-fall season. The bulkhead behind the hoppers was added when the unit was converted for departmental use and the doorway within it leads to one of the unit's two route learning saloons.

Class 121 unit 55025 was converted into a route learning and Sandite unit back in NSE days, when it was given departmental number 977859. It was later rebuilt as a video survey unit for Railtrack and Balfour Beatty and given unit number 960011. From time to time the unit visited Aylesbury for maintenance and it is seen here in the branch siding on 4 July 2004.

With much of the camera equipment now removed and in a fresh coat of unbranded Railtrack livery, 960011 passes Woodhouse working the 5Z21 09.28 Derby–Derby via Woodhouse, Sheffield, Toton, Leicester and Bedford on 8 March 2007. The unit became known as *Pandora* and carries the name on the former headcode panel. (Chris Booth)

Another NSE Class 121 route learning and Sandite conversion was 960014 (977860), the former 55028. After a period stored out of use at Reading, it was purchased by Chiltern Railways in 2003 and moved to Aylesbury, where it is seen stored in the branch siding on 11 August 2003. The following year it was repaired, repainted in BR blue and grey and put into service as a route learner. Its first duties were driver training runs on the Hatton–Stratford-upon-Avon line in preparation for the transfer of Stratford services from First Great Western Link to Chiltern Railways.

Inside Aylesbury TMD, 960014 is under repair in the company of 121020 on 6 March 2005. As well as meeting the route learning needs of Chiltern Railways, the unit is often hired out to other TOCs and organisations for driver training or test train work.

No. 960014 is on a Sandite working at Marylebone with 960010 on a rather wet 30 November 2008. The unit had previously been used for Wrexham & Shropshire driver training and was still carrying vinyls to this effect. No. 960014 did not have its Sandite equipment restored to working order when it was reactivated by Chiltern Railways so 960010 was providing all the Sandite. However, 960014 was added to the formation to provide insurance against breakdowns.

No. 960014 at Aylesbury TMD after a test run on 1 June 2014. It is fitted with equipment to trial a highly accurate positioning system that may one day replace track circuits and axle counters.

No. 960021 (977723/55021) was yet another of the former NSE route learning and Sandite conversions and was the only Class 121 unit to receive the tan, white and grey Railtrack livery. Operated by this time by Chiltern Railways, it is seen here at Aylesbury with 165011 on 16 September 2002. This unit was fitted with two additional headlights beneath the centre windscreen.

No. 960021 is stabled in the Down bay platform at Banbury on 3 November 2004. It was being used for route learning over the Hatton–Stratford-upon-Avon line prior to the transfer of this service from First Great Western Link to Chiltern Railways. The usual unit for this job, 960014, was unavailable on this day. Prominent behind the centre windscreen is the back of the Sandite control console.

Aylesbury's branch siding has almost constantly hosted heritage DMUs for the past decade and a half but this line-up on 23 July 2001 has to be the most varied and impressive yet recorded. There are no less than seven Class 121 and 122 'bubble cars' present: Railtrack tan 960021 (55021), Silverlink 121027 (55027), Railtrack tan 960015 (55019), maroon 960010 (55024), NSE 121031 (55031), NSE 960013 (55030) and NSE 960014 (55022). The two ex-Silverlink units (121027 and 121031) had been purchased by Chiltern Railways as part of a manoeuvre that would allow them to retain the departmental Sandite units they operated on behalf of Railtrack. Railtrack wanted two units for conversion into emergency vehicles for the Severn Tunnel, and they proposed to use two of the Aylesbury-based Sandite units. Chiltern brokered a deal whereby they would purchase two of the redundant Silverlink 121s and hand them over to Railtrack in exchange for the Sandite units, which would henceforth belong to Chiltern.

Former NSE route learning and Sandite unit 960013 (977866/55030) was purchased by Chiltern Railways as a source of spares and never turned a wheel under its own power while at Aylesbury. After being stored here on the branch siding for over five years, it was eventually scrapped along with 960021 in 2011. It is seen on 15 September 2008.

Nos. 960021 and 960013 are prepared for scrapping at Aylesbury TMD on 5 March 2011. No. 960021 is being stripped of its Sandite equipment and 960013 is having its residual asbestos removed, hence the temporary polythene 'air lock' next to the cab.

The underframe of 960021 in the process of being cut up for scrap. Note the 'antler' exhaust pipes sagging forlornly at the far end. Sister unit 960013 has already been dispatched, one of its bogies being visible on the left in this photograph from 3 April 2011.

No. 960010 (977858/55024) was the first route learning and Sandite unit to be allocated to Aylesbury, arriving at the Buckinghamshire depot in NSE days. After privatisation it became the property of Railtrack but it was operated and maintained by Chiltern Railways, who took the decision to paint it in BR maroon rather than Railtrack tan, white and grey. The only concession to the official livery specification was the 'RAILTRACK – CLEARING THE WAY' lettering on the bodysides. It is seen here departing from Great Missenden with a driver training run on 6 March 2003.

No. 960010 undergoes a repaint inside Aylesbury TMD on 18 July 2004. The instruction from Network Rail for a repaint into their standard yellow livery was somehow 'misunderstood' and the unit received a new coat of maroon, looking even better this time for the loss of the Railtrack lettering. The unit later passed from Network Rail to Chiltern Railways' ownership.

The rare sight of a DMU shunting loco-hauled coaches. Pending the arrival of a dedicated diesel shunter, Chiltern Railways made use of Class 121 DMUs to shunt Wrexham & Shropshire loco-hauled sets when they arrived for tyre turning at Aylesbury. No. 960010 is dragging DVT 82304 and a buffet car out of the wheel lathe on 9 October 2009. The single exhaust plume gives away the fact that only one of 960010's engines is working.

By 2011 960010 had been withdrawn from service and it is seen here stored in the branch siding at Aylesbury on 20 June of that year. It was subsequently sold for a nominal £1 to the Chinnor & Princes Risborough Railway, where it has joined sister unit 55023.

The interior of the no. 1 end of 960010, showing part of the route learning saloon and cab. The curved shelf on the left would have been equipped with a kettle when the unit was in service.

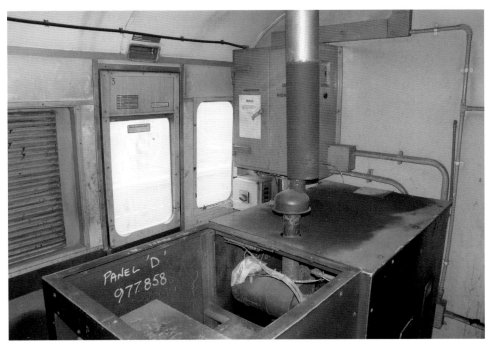

The generator inside the central compartment of 960010. It provided the power for the Sandite pumps.

Notwithstanding the major structural changes to the vehicle, the former guard's compartment on 960010 has survived largely unchanged, including the rack of pigeon holes. Despite this, under normal circumstances there is no requirement for a guard to be present when the unit is performing its departmental duties.

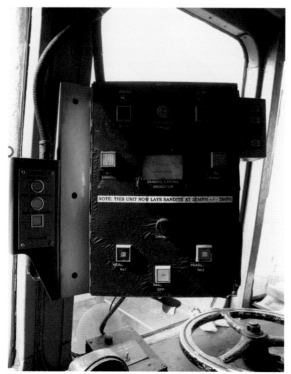

One of the Sandite control consoles inside 960010. On the left is the TPWS panel.

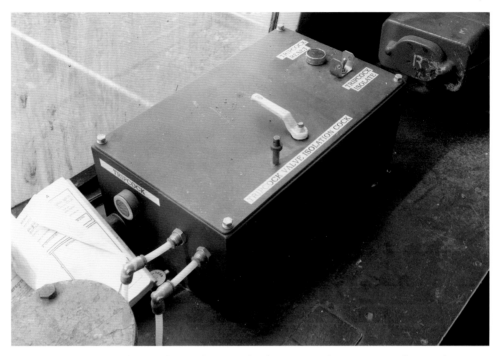

In order to work over the London Underground infrastructure between Amersham and Harrow on the Hill, Chiltern's departmental DMUs have to be fitted with tripcocks. This box on the secondman's side of the cab contains the control equipment as well as the reset switch and isolation cock for the system.

No. 960301 is a three-car, former Class 117 DMU which was converted into a water jetting and Sandite unit for Chiltern Railways. It is seen here on Aylesbury TMD's 'A' road, after having been fuelled on 25 March 2007. When its water tanks are fully filled, the unit is very heavy, and drivers have to take care to brake early and anticipate cautionary signals.

With its water tanks freshly filled, 960301 comes off Aylesbury TMD ready for another day of service on 2 November 2009. The two driving cars (977987/51372 and 977988/51413) came from the former Scotrail unit 117306. In recent years its usage has been restricted to the Aylesbury–Amersham–Marylebone line, under contract to London Underground Ltd, who own the lion's share of the infrastructure on that route. Only water jetting is required for these duties so 960301's Sandite facility is no longer operational.

When first delivered in 2003, 960301 was a two-car unit. A need for additional water capacity was identified and in 2005 a centre car was added. A trailer car would have left the unit underpowered so an Intermediate Motor Vehicle (IMV) was created by removing the cab from Class 117 DMS 51375 (ex-Silverlink unit 117724) and fitting it with a standard gangway end. It is seen here from the former cab end before it was inserted into 960301.

At the non-driving end of 977988 is this large diesel engine, which drives the water jetting pumps. Behind the engine is one of the many water tanks that are distributed across the remainder of the unit. Note how all the interior wall and roof panelling has been removed in order to save weight.

As it works over London Underground metals, 960301 is fitted with LU's tripcock system. The wooden beam on the leading bogie holds the mechanism and beneath the right-hand bolt in the beam, the tripcock arm itself can be seen. When an LU signal is showing danger, a raised trainstop lifts up. If a train fails to stop at the signal the trainstop hits the tripcock and engages the vehicle's brakes. Also visible here is one of the now redundant Sandite nozzles.

The water jetting process blasts the railheads with high-pressure water and cleans off any accumulated leaf contamination. It is a messy process, and 960301 is permanently grimy. This unit is known as the 'Green Goddess' among Chiltern staff, as this unofficial embellishment on the bodyside dirt demonstrates.

Anatomy of a DMU

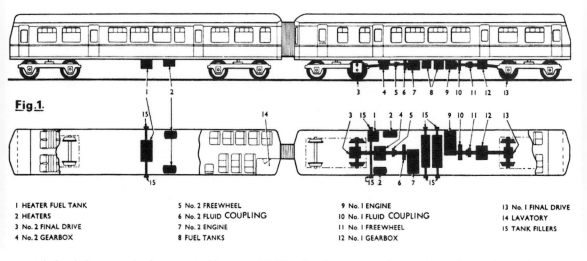

Fig.1.

1 HEATER FUEL TANK	5 No. 2 FREEWHEEL	9 No. 1 ENGINE	13 No. 1 FINAL DRIVE
2 HEATERS	6 No. 2 FLUID COUPLING	10 No. 1 FLUID COUPLING	14 LAVATORY
3 No. 2 FINAL DRIVE	7 No. 2 ENGINE	11 No. 1 FREEWHEEL	15 TANK FILLERS
4 No. 2 GEARBOX	8 FUEL TANKS	12 No. 1 GEARBOX	

A behind-the-scenes look at a typical heritage DMU in detail, concentrating on things that are beneath the solebar, inside the cab or otherwise beyond the view of the passenger. The diagram above comes from an old BR instruction manual and shows the general layout of a heritage DMU. Each engine is connected via a fluid coupling and freewheel to an epicyclic gearbox. A cardan shaft connects each gearbox to a final drive unit on the innermost axle of the adjacent bogie.

Also illustrated in this section are the main elements of a typical heritage DMU cab. The cab controls are broadly consistent across most heritage DMU types but the exact layout of those controls can vary from class to class. Illustrated here is the 'Derby' type cab shared by classes 117, 121 and 122 with minor class-specific variations. The Class 101 cab is somewhat different but is illustrated in the First North Western section of this book.

Above: Detail of a BUT 150 hp 'L' type (Leyland) engine. The majority of heritage DMU types use two of these engines per motor vehicle (55020).

Right: Engine start/stop buttons and local throttle control (55020).

Left: The fluid coupling which replaces a conventional clutch and serves as a flywheel to even out irregularities in the power output (55020).

Below: Four-speed, air-operated epicyclic gearbox. The freewheel unit is partially obscured on the right (55020).

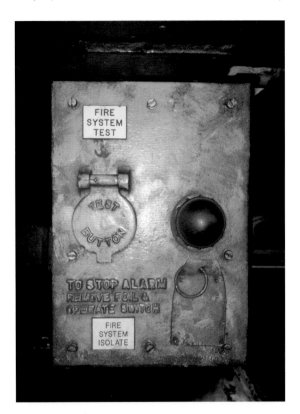

Engine fire panel with test button, fire alarm indication light and double pole switch, which stops the alarm and prevents the engine from being restarted (55034).

Battery isolation switch on the electrical box (55020).

Battery box with the covers removed (51411).

The main EP valves with the cover removed. The bottom row control the throttle and direction controller functions; the top row control the four gears (55020).

The lighting control box, behind which is the vacuum brake reservoir (55034).

Control air reservoirs for the EP system, with one of the fuel tanks behind (55034).

DMU bogies with a vacuum brake cylinder prominently visible (55021).

A final drive unit on a spare driving wheelset.

British Railways Western Region coupling weight (55020).

A Pressed Steel Co. works plate, attached to the vehicle's solebar (55034).

Buffer beam detail, showing the coupling and the twin pipes for the Gresham & Craven DMU quick release vacuum brake. The right-hand hose is the Train Pipe, which contains a vacuum of up to 21"Hg. The left-hand hose is the Release Pipe, which contains a vacuum of up to 28"Hg (51532).

A standard DMU marker/tail light fitting with the cover open (marker light bulb not present). These have been replaced with LED units on the Chiltern and ATW passenger 121s (977988).

The guard's emergency brake valve (left) and the 'toasting fork' tool (right), which is used for isolating the final drives and manually activating the engine stop solenoid – a requirement in certain fault conditions (55034).

The driver's tools of the trade for a heritage DMU. At the top is the Gresham & Craven vacuum brake handle. On the left is the reversing handle, known colloquially as the 'spoon'. At the centre is the FG1 control circuit key.

A typical heritage DMU 'Derby'-type driving desk, shared by Classes 117, 121 and 122, with minor class-specific variations. The throttle controller is on the left, with the gear selector on the right. The vacuum brake is just out of shot on the right. The four gauges from left to right are: speedometer, tachometer, vacuum brake gauge and control air gauge. At the centre of the desk is the control circuit key slot. The Class 101 cab is somewhat different, but is illustrated in the First North Western section of this book.

Above left: The left-hand cabside controls, from top to bottom: buzzer speaker and vacuum brake low topside warning lights (121020/34 only), engine and air/axle confirmation lights, engine start/stop buttons and door release buttons (121020/34 only).

Above right: Modern additions to a DMU cab. From top to bottom, along the right of the driver's windscreen: PIS input panel, GSM-R radio, TPWS panel, GSM-R radio handset, door close/release buttons, DRA (55020).

Acknowledgements

Photographic contributions are gratefully acknowledged from the following: Lewis Bevan, Chris Booth, Rowan Cranshaw, Dave Evans, Ian Buck, Martin Hawkes, Antony Henley (http://www.hentis-rail.co.uk) and Charlie Hulme (http://nwrail.org.uk/).

The driver's Gresham & Craven brake controller and the two-tone horn valve below (55020).

Opposite bottom: The emergency brake valve situated on the secondman's side of the driving desk (55020).

Key to Abbreviations

ATC	Automatic Train Control	DSD	Driver's Safety Device
ATW	Arriva Trains Wales	DVT	Driving Van Trailer
AWS	Automatic Warning System	ECS	Empty Coaching Stock
BR	British Rail	EMU	Electric Multiple Unit
CCTV	Closed Circuit Television	EP	Electro-Pneumatic
CSR	Cab Secure Radio	FGE	First Great Eastern
DEMU	Diesel Electric Multiple Unit	FNW	First North Western
DMBS	Driving Motor Brake Standard	GSM-R	Group Standard Mobile – Railway
DMBSO	Driving Motor Brake Standard Open	IMV	Intermediate Motor Vehicle
DMS	Driving Motor Standard	LED	Light Emitting Diode
DMSL	Driving Motor Standard with Lavatory	LU	London Underground
		NRN	National Radio Network
		NSE	Network South East
DMU	Diesel Multiple Unit	PIS	Passenger Information System
DMSL	Driving Motor Standard with Lavatory	TCsoL	Trailer Composite Semi-Open with Lavatory
DTCsoL	Driving Trailer Composite Semi-Open with Lavatory	TOC	Train Operating Company
		TOU	Train Operating Unit
DTSO	Driving Trailer Standard Open	TPWS	Train Protection & Warning System
DTSL	Driving Trailer Standard with Lavatory	TSL	Trailer Standard with Lavatory
DTSOL	Driving Trailer Standard Open with Lavatory	TSO	Trailer Standard Open
		TSOL	Trailer Standard Open with Lavatory
DRA	Driver Reminder Appliance		